I0820073

When scientists get things right,
science leaps forward.
When they get things wrong,
they learn from their mistakes,
and science still marches forward.
But sometimes,
even when scientists get it right,
nobody pays attention.

Then,
for a while,
science stands still.

When
Science

Written by
Shruthi Rao

Illustrated by
Srinidhi Srinivasan

Stood Still

How S. Chandrasekhar Predicted the Existence of Black Holes

Margaret K. McElderry Books
New York Amsterdam/Antwerp London Toronto Sydney/Melbourne New Delhi

For Subrahmanyan Chandrasekhar, math wasn't made up of problems, but of solutions.

Sitting by himself among his grandfather's books in the South Indian city of Madras, Chandra devoured books on advanced math.

He often biked to the beach to sit under the twinkling stars, those big balls of gases made of a gazillion atoms swirling around, gravity squishing those atoms together so hard that they combined to give out heat and light.

He gazed at the stars and dreamed of uncovering their secrets.

When he was eighteen,
an idea from here . .

a fact from there . . .

gave him a new understanding
of how gases behave in stars.

The young scientist presented his work to a gathering of top Indian scientists to applause and appreciation.

The government of India gave him a scholarship to go to England, to study astrophysics—the science of stars.

As the ship sailed across the Indian Ocean, Chandra sat in the sun on the deck of the ship, thinking deeply about the life and death of stars. His pencil danced across a piece of paper, filling it with numbers.

With an idea from here
and an equation from there
coming together on his page,

BOOM!

an odd, yet incredible new idea exploded in his head.

Chandra had just grasped the idea that a dying star could result in something mind-boggling—an object that would decades later come to be known as a "black hole."

What an astonishing idea! How could this be?

But his math was right.

It simply had to be.

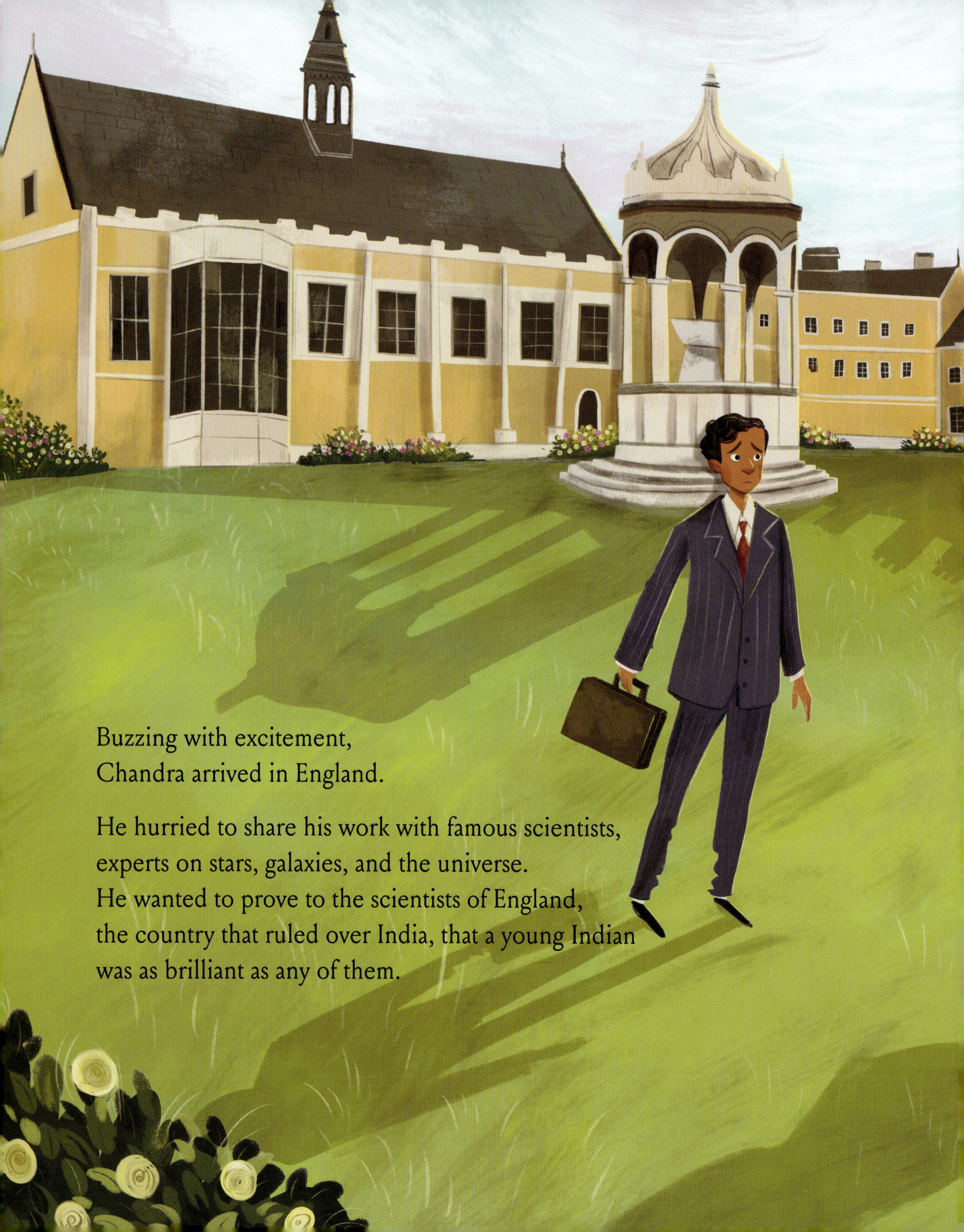

Buzzing with excitement,
Chandra arrived in England.

He hurried to share his work with famous scientists,
experts on stars, galaxies, and the universe.
He wanted to prove to the scientists of England,
the country that ruled over India, that a young Indian
was as brilliant as any of them.

But . . .

They ignored him.

Everyone, except Sir Arthur Eddington. One of the very scientists whose books Chandra had grown up reading.

Eddington arranged for Chandra to present his exciting new ideas in front of an important audience.

Chandra stood up and spoke with confidence.

They all thought dying stars quietly cooled down into planet-sized balls, didn't they?

Well, not all stars!

Some stars, bigger than our sun, end their lives in a massive explosion, like gigantic fireworks in space!

And then, they shrink into a teeny-tiny dot, and . . .

disappear!

Odd, yet incredible, this little discovery could lead to big things. They could all work together to find secrets of the universe they never knew before, couldn't they?

Chandra sat down and waited—

for their applause,
for their wonder,
for their excitement.
But . . .

Eddington stood up, and . . .

He tore down Chandra's ideas,
and joked about them.

The audience roared with laughter at this audacious young man from across the seas with ridiculous ideas in his head.

Whoever heard of a star exploding,
and then disappearing!

They were sure Chandra's math was all wrong.

Chandra felt
betrayed,
shocked,
confused.

He knew he was right.
Why couldn't they see it?

Was it because Eddington was such a brilliant star in his field that they couldn't even see the truth in his dazzling light?

Or did they just not want to take Chandra's side against Eddington?

For years, Chandra tried to get them all as excited as he was . . .

in vain.

Finally, he'd had enough.
He couldn't waste any more time.
After all, there were many more discoveries to be made.

The stars would still keep doing what he knew they did—glowing, shining, exploding, disappearing—whether anybody believed him or not.

He knew, someday, the world would see the truth of the universe.

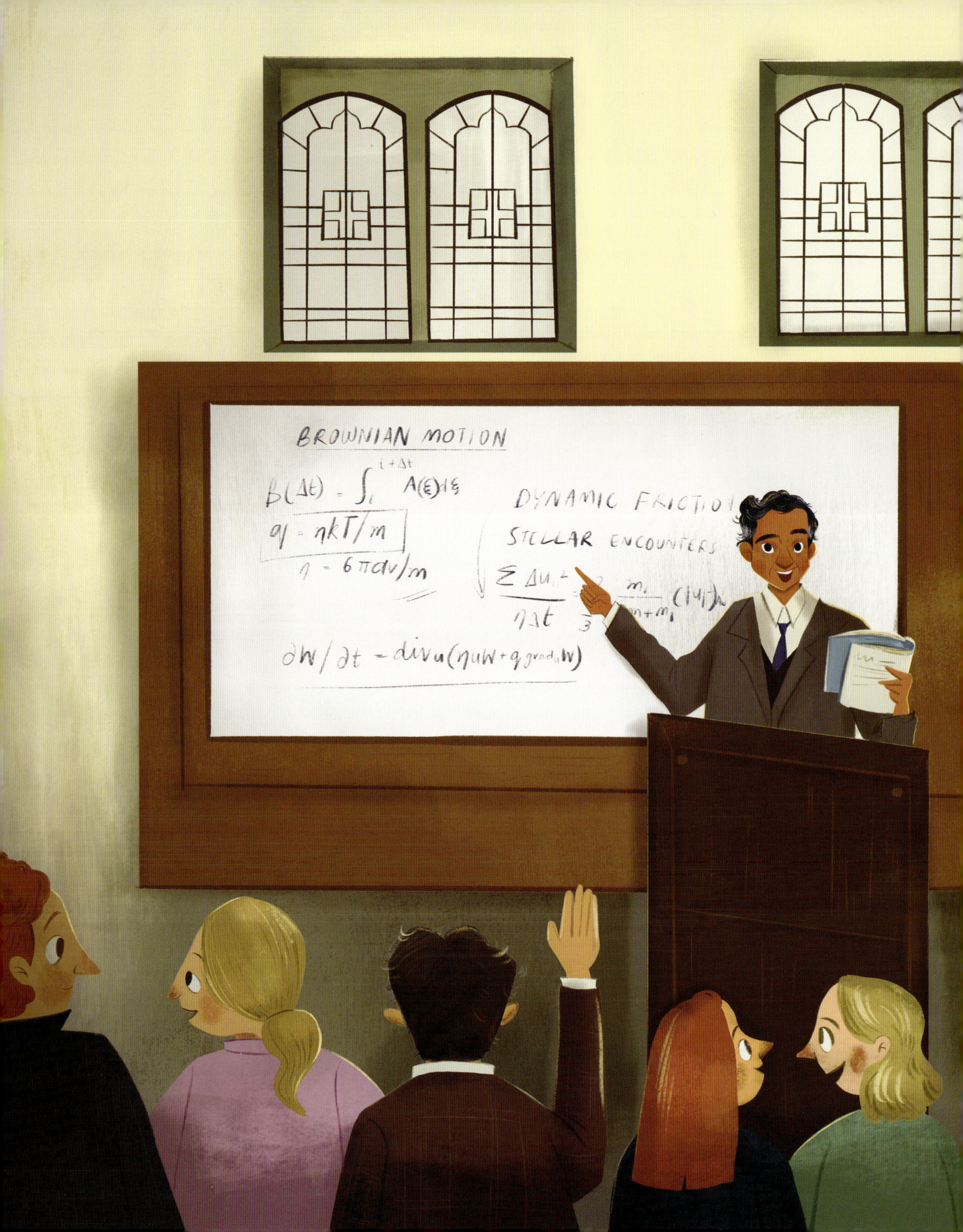
BROWNIAN MOTION
DYNAMIC FRICTIO
STELLAR ENCOUNTERS

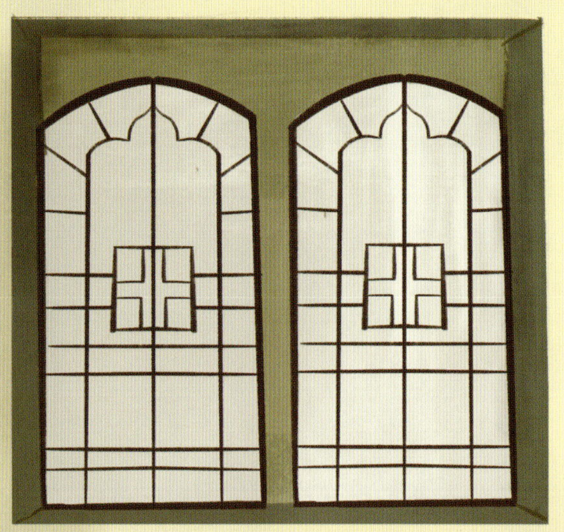

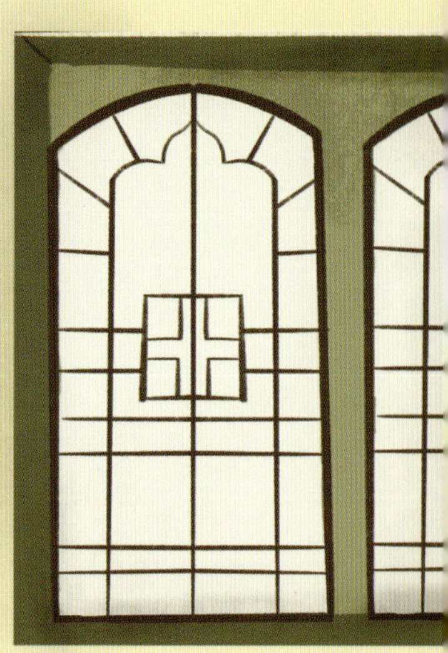

Chandra went to live and work in America.
Through math, he discovered other secrets of the universe.
He taught younger scientists.
He wrote books on astrophysics.
He won awards.

Chandra moved on.
But the science of stars stood still.

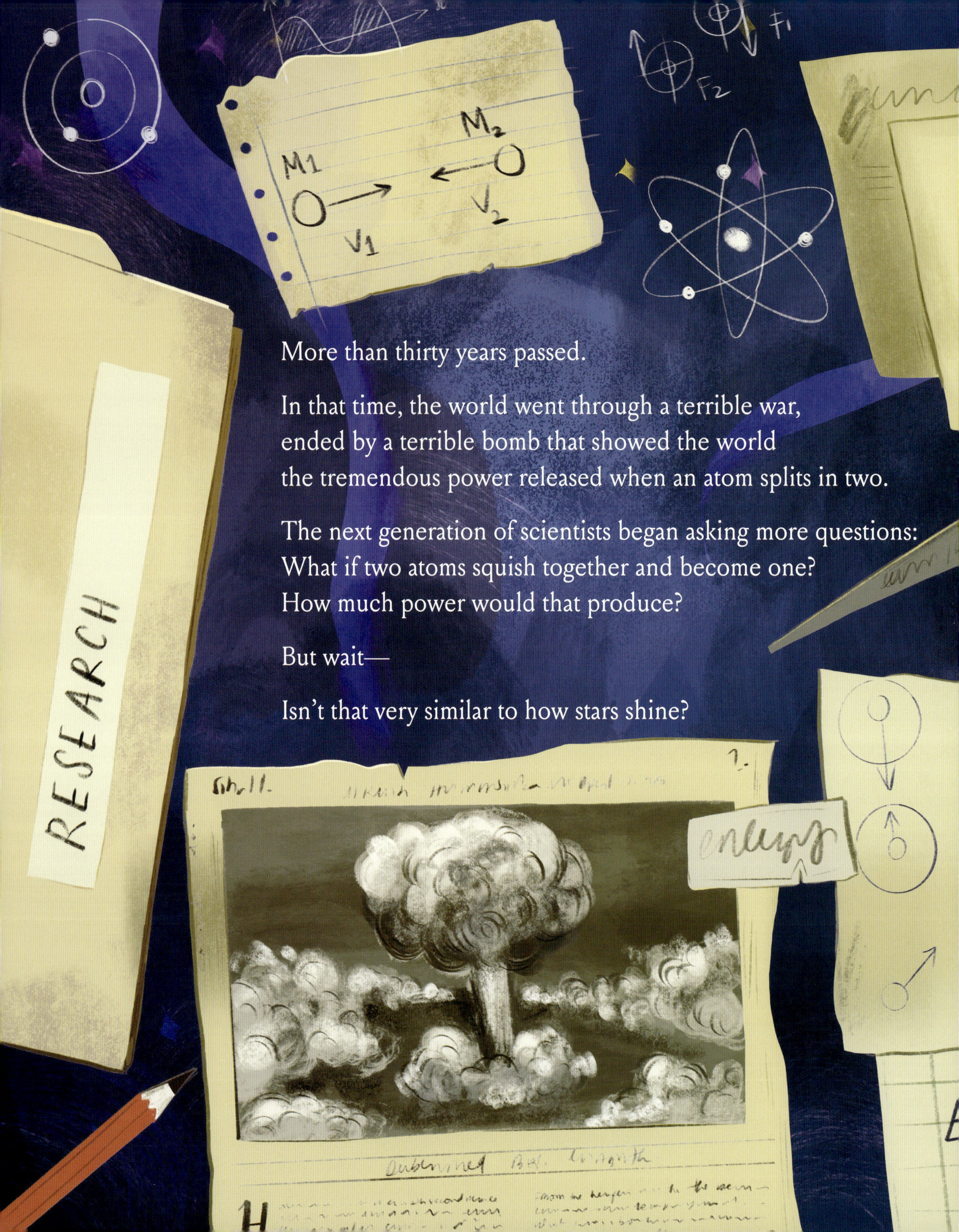

More than thirty years passed.

In that time, the world went through a terrible war,
ended by a terrible bomb that showed the world
the tremendous power released when an atom splits in two.

The next generation of scientists began asking more questions:
What if two atoms squish together and become one?
How much power would that produce?

But wait—

Isn't that very similar to how stars shine?

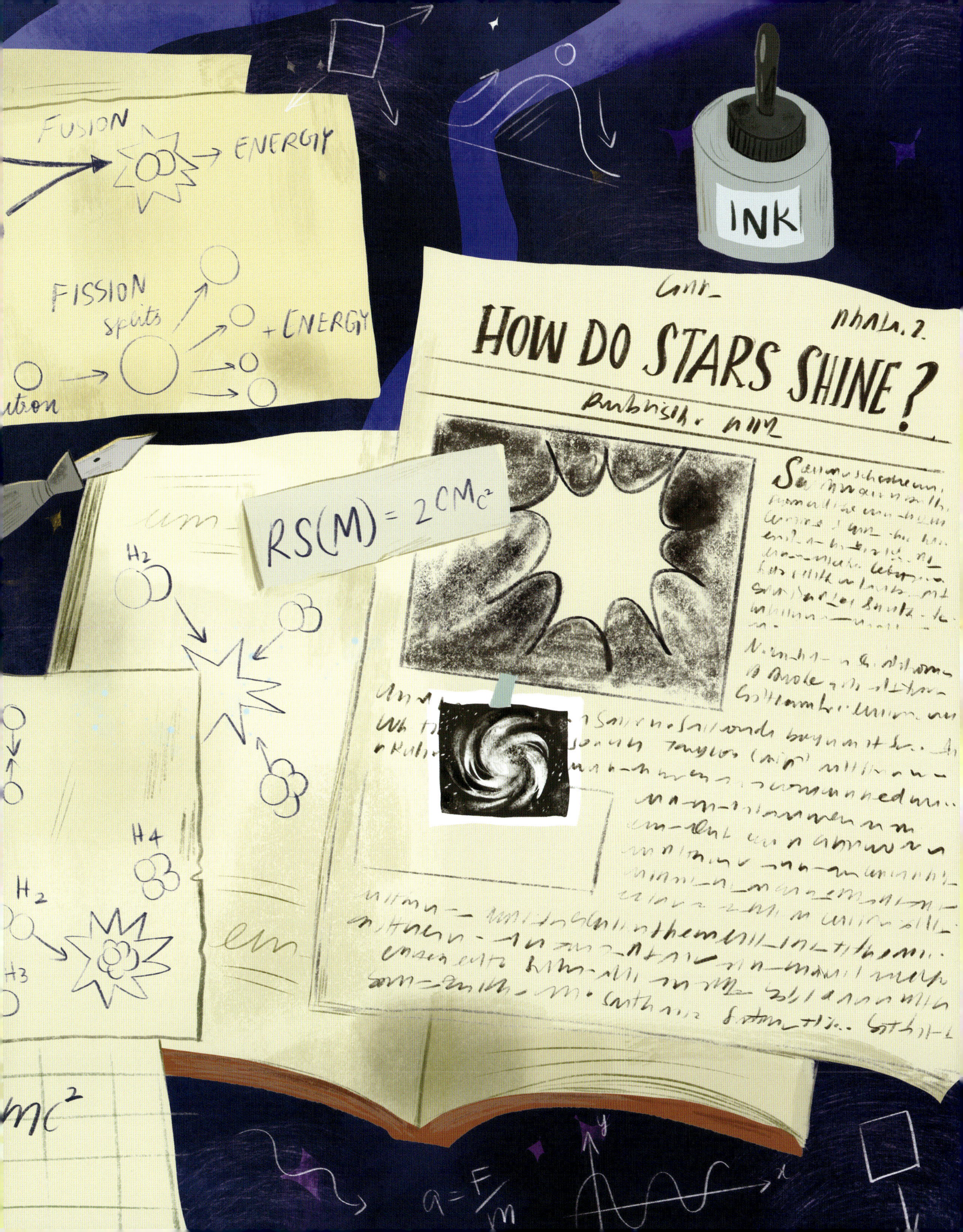
FUSION
ENERGY
FISSION
splits
+ ENERGY
INK
HOW DO STARS SHINE?
RS(M) = 2CMc²
H2
H4
H2
H3
a = F/m

All eyes turned again toward the skies, and the scientists found their answers, and something more—exploding stars that shrank and seemed to disappear! Exactly what Chandra had insisted all those years ago.

But this time, the world listened, and understood what Chandra had meant: this could lead to big things!

Many scientists, including Chandra, put their heads together.

With the help of smart new computers, measuring instruments, and powerful telescopes, they looked deep into the dark, mysterious expanse of space and discovered objects they never knew existed, including the most exciting one of all—a crevice into which shrinking stars slip and disappear—

a black hole!

This discovery has led to a better understanding
of the universe,
of how it works,
and how it began!

An understanding that came decades late because the world ignored a brilliant young scientist.

Finally, about fifty years after Chandra first talked about stars disappearing, he won the Nobel Prize for his work. More than the recognition, what mattered to him was that, finally, the world saw the truth.

And once again,
science was leaping forward.

A Closer Look

Subrahmanyan Chandrasekhar

[soo-BRUH-mun-yun CHUN-druh-SHAY-khur] (1910–1995), popularly known as Chandra [CHUN-druh], was an Indian-born American astrophysicist.

Chandra was a theoretical astrophysicist—a scientist who uses math to explain and predict how stars behave.

He was born in Lahore (in present-day Pakistan) and later moved to Madras (now Chennai), in India, where his family was originally from. Chandra grew up in a house surrounded by books. He was caught up in the patriotism that accompanied the struggle for freedom from British rule, and he felt that by doing well in science, he would prove to the world what Indians were capable of. He was inspired by his uncle, the famous scientist C. V. Raman, who won the 1930 Nobel Prize in Physics.

During his student days in India, Chandra was already well-known for his studies in astrophysics and published papers in international journals. Recognizing his potential, the government of India gave him a scholarship to study in England.

He was just nineteen when, on the voyage to England, Chandra made his groundbreaking discovery. Einstein's special theory of relativity explains how speed affects mass, time, and space, especially the speed of light. Chandra thought of applying this theory to stars, because particles inside stars travel at speeds close to that of light. That is when his calculations suggested that some dying stars behave differently.

Chandra completed his PhD from Cambridge University in 1933 and got a fellowship at Trinity College. After the rejection and ridicule he faced because of his theory, he decided to move on. He left for the University of Chicago in 1937, where over the next several decades he carried out research in other fields of astrophysics, such as energy transfer by radiation in the atmospheres of stars, and on convection on the solar surface. He received several awards and recognition: the Gold Medal of the Royal Astronomical Society, the Royal and Copley Medals of the Royal Society, the National Medal of Science, and many other honors. He became a US citizen in 1953.

In the 1970s, by which time the world had found evidence confirming that Chandra's theory of dying stars was true, he went back to working on the fate of the stars at the end of their lives.

Chandra was a popular teacher and many of his students went on to win Nobel prizes themselves. He wrote several books and was the editor of the *Astrophysical Journal* for nineteen years.

Subrahmanyan Chandrasekhar died in 1995. According to his wishes, his Nobel Prize money was used to create a fellowship for promising young astrophysicists at the University of Chicago.

The Life and Death of Stars

A star is born from a giant cloud of gas and dust called a nebula. The particles begin to gather into a gigantic ball due to the gravitatioal attraction between them. The closer they get, the greater the attraction between them. The inside of the ball, called the core, gets squeezed so hard that hydrogen atoms begin to combine to form helium (in a process called fusion—see below), releasing energy, and a star is formed.

Within the star, the inward force of gravity in the core is balanced by the outward forces of the energy released by fusion. In this state, the star burns steadily for billions of years. Our Sun is in this stable state now and is said to be in "the main sequence."

Once all the hydrogen gets used up in the core of the star, gravity wins and begins compressing the star.

What happens next depends on the size of the star.

In a smaller star, such as our Sun, the compression of the core leads to higher temperatures and more energy generation in the outer layers of the star. The star becomes very big and very bright, and is called a red giant. The gases in the outer layers get dispersed into space. The inner core remains and is called a white dwarf. It burns with a faint light until it completely dies out.

A larger star, however, goes through a more dramatic phase. It blows out its outer cover in a massive explosion called a supernova, and the remains start collapsing inward. It becomes a dense neutron star, as large as a city, but with as much mass as thousands of suns.

A star that is even larger has so much gravity that after the supernova explosion, it collapses inward in a whoosh until an immense amount of mass is contained in a tiny point called a singularity. The gravitational pull of this point is so strong that it sucks up everything around it, even light. This is a black hole.

At the meeting at the Royal Astronomical Society in England in 1935, at which he and Eddington clashed, Chandrasekhar not only told his audience that a large star does not end up as a white dwarf, but also that there was a threshold beyond which a white dwarf cannot exist. A white dwarf with a mass less than 1.4 times the mass of the Sun will fade slowly away and die with a whimper. But beyond this threshold, the white dwarf collapses (into what we now know are neutron stars or black holes). This threshold has come to be known as the "Chandrasekhar Limit."

A Scientific Controversy

Arthur Stanley Eddington (1882–1944) was an English astrophysicist. Eddington shot to fame when he proved Albert Einstein's general theory of relativity. Einstein had said that large objects in space, like the Sun, bend the path of light. In 1919, Eddington went on an expedition to West Africa to verify this during a solar eclipse. His measurements proved Einstein right, and both Einstein and Eddington became well known all over the world.

Nobody knows for sure why Eddington encouraged Chandra at first and then attacked him publicly. The obvious explanation would be that Chandra's theory proved Eddington, who thought all stars die with a whimper, wrong. But scientists and historians have remained puzzled over Eddington's extreme reaction. It is also surprising that Chandra, being young, new, and an immigrant, did not get crushed by this experience, but held on to his convictions and moved on.

Bombs and Stars

Chandra wasn't the only person to try to dig into the fate of stars. The American scientist J. Robert Oppenheimer and the Russian scientist Lev Landau studied stars too and got similar results. But those studies did not progress further at that time.

In the 1950s and '60s, the nuclear arms race was heating up, and scientists were trying to make bombs more powerful than the atomic bomb that the United States dropped on Hiroshima and Nagasaki in 1945. In an atomic bomb, energy is released by fission, an atom splitting in two. Scientists were studying the hydrogen bomb, which works by fusion, in which two atoms combine to form another atom. Fusion produces much more energy than fission does. To understand fusion better, scientists turned their attention to the stars, which produce energy by fusion. In the process, they confirmed Chandra's theory.

Today, scientists study the stars and space to understand how we humans and our planet came to be. We are able to use that knowledge to improve our lives, and to see if humans have a future beyond Earth.

Trivia

NASA's Chandra X-ray Observatory is named after Subrahmanyan Chandrasekhar. It is a space telescope, the world's most powerful X-ray telescope. It allows scientists to get X-ray images of space to help understand how the universe was born and evolved. It is part of NASA's fleet of "Great Observatories" along with the Hubble Space Telescope, the Spitzer Space Telescope, and the Compton Gamma Ray Observatory.

Partial Time Line of Black Holes

1915—Albert Einstein adds to his theory of relativity that he first proposed in 1905. These additions provided the basis for the calculations that Chandra did.

1931—S. Chandrasekhar publishes a paper proving that large/heavy stars end their lives in a very different way than everybody thought.

1939—J. Robert Oppenheimer, the man behind the atomic bomb, concludes that a large star ends its life by collapsing.

1963—Maarten Schmidt discovers a quasar, one of the most powerful objects in the universe. Scientists later find that quasars are powered by black holes.

1967—John Wheeler coins the term "black holes."

1971—Scientists identify a black hole for the first time. They name it Cygnus X-1.

2019—NASA telescopes capture an image of a black hole for the first time.

Sources

A Black Hole Is Not a Hole by Carolyn Cinami DeCristofano, Charlesbridge, 2012.

Chandra: A Biography of S. Chandrasekhar by Kameshwar C. Wali, University of Chicago Press, 1990.

Empire of the Stars: Obsession, Friendship, and Betrayal in the Quest for Black Holes by Arthur I. Miller, Mariner Books, 2005.

Oral History Interviews: American Institute of Physics—S. Chandrasekhar interviewed by Kevin Krisciunas; Location: University of Chicago; Date: October 1987 www.aip.org/history-programs/niels-bohr-library/oral-histories/4551-1

Acknowledgments

Thank you to the people who worked on this book: Nicole Fiorica, Lauren Rille, Bridget Madsen, Tatyana Rosalia, and the rest of the team at Simon & Schuster.—S. R.

MARGARET K. McELDERRY BOOKS • An imprint of Simon & Schuster Children's Publishing Division • 1230 Avenue of the Americas, New York, New York 10020 • For more than 100 years, Simon & Schuster has championed authors and the stories they create. By respecting the copyright of an author's intellectual property, you enable Simon & Schuster and the author to continue publishing exceptional books for years to come. We thank you for supporting the author's copyright by purchasing an authorized edition of this book. • No amount of this book may be reproduced or stored in any format, nor may it be uploaded to any website, database, language-learning model, or other repository, retrieval, or artificial intelligence system without express permission. All rights reserved. Inquiries may be directed to Simon & Schuster, 1230 Avenue of the Americas, New York, NY 10020 or permissions@simonandschuster.com. • Text © 2025 by Shruthi Rao • Illustration © 2025 by Srinidhi Srinivasan • Book design by Lauren Rille • All rights reserved, including the right of reproduction in whole or in part in any form. • MARGARET K. McELDERRY BOOKS is a trademark of Simon & Schuster, LLC. • For information about special discounts for bulk purchases, please contact Simon & Schuster Special Sales at 1-866-506-1949 or business@simonandschuster.com. • Simon & Schuster strongly believes in freedom of expression and stands against censorship in all its forms. For more information, visit BooksBelong.com. • The Simon & Schuster Speakers Bureau can bring authors to your live event. For more information or to book an event, contact the Simon & Schuster Speakers Bureau at 1-866-248-3049 or visit our website at www.simonspeakers.com. • The text for this book was set in Louize. • The illustrations for this book were rendered on paper, then finished digitally.
Manufactured in China
0625 SCP
First Edition
10 9 8 7 6 5 4 3 2 1
Library of Congress Cataloging-in-Publication Data • Names: Rao, Shruthi, author. | Srinivasan, Srinidhi, illustrator. • Title: When science stood still : how S. Chandrasekhar predicted the existence of black holes / written by Shruthi Rao ; illustrated by Srinidhi Srinivasan • Description: First edition. | New York : Margaret K. McElderry Books, [2025] | Includes bibliographical references. | Audience: Ages 4 to 8 | Audience: Grades 2–3 | Summary: "When physicist Subrahmanyan Chandrasekhar left his home country of India, his discovery of the existence of black holes was initially met with ridicule, but soon enough, his peers and the whole world would see that his prediction was right all along."—Provided by publisher. • Identifiers: LCCN 2024013658 (print) | LCCN 2024013659 (ebook) | ISBN 9781665949965 (hardcover) | ISBN 9781665949972 (ebook) • Subjects: LCSH: Chandrasekhar, S. (Subrahmanyan), 1910–1995. | Astrophysicists—India—Biography. | Astrophysicists—United States—Biography. | Black holes (Astronomy) • Classification: LCC QB460.72.C48 R36 2025 (print) | LCC QB460.72.C48 (ebook) | DDC 520.92 [B]—dc23/eng/20240508 • LC record available at https://lccn.loc.gov/2024013658 • LC ebook record available at https://lccn.loc.gov/2024013659

To my parents,
Nagraj Rao and Brinda Rao.
Whether I stood still or leaped forward,
I knew I'd be okay. Thank you.
—S. R.

To Sandy,
for lifting me high
to reach for the stars
—S. S.